Anny Quintana Portal
Gretel Dorta Preciado
Yoidel Martínez Morales

Mathematics teaching-learning process

Anny Quintana Portal
Gretel Dorta Preciado
Yoidel Martínez Morales

Mathematics teaching-learning process

System of activities for the development of calculus skills

ScienciaScripts

Imprint
Any brand names and product names mentioned in this book are subject to trademark, brand or patent protection and are trademarks or registered trademarks of their respective holders. The use of brand names, product names, common names, trade names, product descriptions etc. even without a particular marking in this work is in no way to be construed to mean that such names may be regarded as unrestricted in respect of trademark and brand protection legislation and could thus be used by anyone.

Cover image: www.ingimage.com

This book is a translation from the original published under ISBN 978-613-9-43844-0.

Publisher:
Sciencia Scripts
is a trademark of
Dodo Books Indian Ocean Ltd. and OmniScriptum S.R.L publishing group

120 High Road, East Finchley, London, N2 9ED, United Kingdom
Str. Armeneasca 28/1, office 1, Chisinau MD-2012, Republic of Moldova, Europe
Printed at: see last page
ISBN: 978-620-8-24061-5

INDEX

INTRODUCTION

The country is going to make a giant leap in education because it is developing a Revolution in education; it is a colossal battle of ideas that the people are waging with the aim of bringing general and integral culture as a guarantee for the Revolution; the Commander in Chief has repeatedly stated that the fundamental role of schools and educators is to achieve a different, fairer society, which evidently implies a new Revolution in education. The process carries out the Battle of Ideas, in education new technologies are used to achieve an efficient result in scientific development, bringing to the classroom new educational transformations, including a television set, a computer, a video, textbooks, Marti's notebooks, dictionaries, etc. The teaching-learning process in the subject of Mathematics in third grade, these transformations are manifested in the appropriate use of the computer and educational software, which are used to link the classes with the subject through the use of software: Mathematics Fairs, The Forms that Surround Us I, The Land of Numbers and others.

The teaching of school mathematics plays an important role in the formation of individuals who are capable of taking on the scientific and technical challenges demanded by current social development. In this sense, it is necessary that pupils at school learn how to learn. By achieving a cognitive dependence in the acquisition of knowledge, the subject can make a significant contribution to these general aims, either by contributing to the development of pupils' general mental capacities or by fostering creativity, imagination and the creation of the habit of discipline, among other things.

In order to understand the meaning of mathematics and its teaching, it is necessary to know its historical development, which shows that mathematical knowledge, arising from the practical needs of man through a long process of abstraction, has a great value for life. The foundations of mathematical science have become essential tools for understanding and transforming the world, and it is therefore necessary for all schoolchildren to learn the basics of this science, so that they can also solve the countless problems that arise in practice and whose solution requires the use of mathematical scaffolding. Mathematical training at school today is becoming even more important for subsequent practical activity, since the role of mathematics in social life is increasing at an especially rapid pace, and the scientific and technical process and the technical complexity of production are placing new demands on the training of new generations.

The position of the socialist educational policy and Marxist-Leninist pedagogy demands a complete pedagogical work, in theory and practice, creating a decisive basis for the teaching of mathematics. However, the lack of motivation to study mathematics and the poor development of skills in this subject are obstacles to the achievement of these aims and constitute difficulties which mathematics teachers have to face systematically in the course of their work. However, there are difficulties presented by third grade children in the basic addition exercises with overcoming in terms of mathematical calculation, due to the difficulties of lacking a good memorisation of the basic addition exercises, as they present deficiencies in working with dexterity in the calculation, as well as lacking skills in it.

- Calculation errors in addition with 10000 limit overrun.
- Poor command of operational order.

In view of the above, the scientific problem is defined as: How to develop calculation skills in basic addition exercises with overrun in third grade students at the Reinaldo León Yera school? The object proposed is: Teaching-learning process in Mathematics in third grade. As a field of action: calculation skills in basic addition exercises with overrun. In order to provide an answer to the problem, the following objective was proposed: To propose a system of activities for the development of calculation skills in basic addition exercises with overrun in third grade students.

The following scientific questions were posed for the development of the research:

1. Systematisation of the theoretical foundations of the teaching-learning process of mathematics and calculation skills in basic addition exercises with overrun in third grade?
2. What is the current state of calculus skills in the basic exercises of addition with overshooting in third graders?
3. What avenues can be used to develop skills in the calculation of addition with surplus in the third grade?
4. How effective is the activity system?

In order to fulfil the proposed objective, the following research tasks were developed:

1. Theoretical foundation of the teaching-learning process of mathematics and calculation skills in basic addition exercises with overshooting.

2. Diagnosis of the current situation of the development of calculation skills in basic addition exercises with overrun in third grade.
3. Design of a system of activities for the development of skills in the calculation of basic addition exercises with overrun.
4. Evaluation of the effectiveness of the system of activities.

The following theoretical methods and techniques are proposed for the following research.

Historical-logical. It allowed to know the phenomenon under study in its background and current trends, which allowed to establish the theoretical bases that supported the research, to reflect logically the essence, necessity and regularity of behaviour in its development of the teaching-learning process in Mathematics in third grade students, from the development of the ideas studied.

Inductive-deductive: To start from the general to the particular, from the specific study of skills in the basic addition exercises with surpassing in third grade.

Analysis-Synthesis: To interpret the theoretical postulates related to the teaching-learning process in Mathematics in third grade students and to synthesise the primary and general issues.

Structural systemic: To relate the parameters of the system of activities, starting from the simple to the more complex.

The empirical level methods used were as follows:

1. Pedagogical test: It was to find out the real state of knowledge and skills that the student possesses in the calculation skills of basic addition exercises with surpassing in the third grade.

in calculus in third grade students at the Reinaldo León Yera school.

2. Observation: To find out the real situation of the skills in the calculation of addition with surplus in the third grade.

3. Interviews with Mathematics teachers revealed the main difficulties that pupils have with regard to the calculation skills of basic addition exercises with overrun in third grade.

4. Surveys of teachers and methodologists of the subject Mathematics to obtain information on students' knowledge of the skills in the calculation of basic addition exercises with overshooting.

5. The experimental method (pre-experiment design) will be used: To validate a system of activities to develop skills in third grade students of the Reinaldo León Yera school.

Statistical methods: It made it possible to compute data obtained through the empirical route.

The population selected for this research is composed of the third grade group at the Reinaldo León Yera school.

A total of seventeen students and the sample matches the population.

The scientific novelty: it consists of the elaboration of a system of activities for the development of skills in the calculation of basic addition exercises with surplus in third grade students.

The practical contribution: a system of activities for the development of numeracy skills in third grade pupils at the Reinaldo León Yera school.

CHAPTER I: THEORETICAL FOUNDATION OF THE MATHEMATICS TEACHING-LEARNING PROCESS.

1.1- Historical overview of the emergence and development of Mathematics

This situation, which has a universal manifestation, is also present in Cuba. For example, in 1984, the situation of mathematics teaching at secondary level was recognised as a problem to be solved in the country. The National Seminar of the Ministry of Education (MINED) spoke of "formalism in the teaching of Mathematics" as a way of signifying the poor contribution of the subject to the meaningful learning of students. The quality of pupils' knowledge of numbers and their abilities and skills in working with them has a great influence on the effectiveness of the subsequent treatment of numeracy with these numbers. Mathematics represents the study of relationships between quantities, magnitudes, properties and the logical operations used to reduce unknown properties. It is a science that is now 2000 years old and although it is now structured and organised. This operation took a very long time. In the past, mathematics was considered as the science of quantity referring to magnitudes (as in geometry), to numbers (as in arithmetic) or to the generalisation of both (as in algebra). Towards the middle of the 19th century, mathematics began to be considered as the science of relations, or as the science that produces necessary conditions. The latter notion encompasses mathematical logic or symbolic logic - the science of using symbols to generate an exact theory of logical deduction or inference based on definitions, axioms, postulates and rules that transform primitive elements into more complex relations and theorems. In fact, mathematics is as old as mankind itself. It is found in prehistoric designs on pottery, textiles and in cave paintings (where evidence of geometric sense and interest in geometric figures can be found). Primitive calculation systems were probably based on the use of the fingers of one or two hands (pay attention as children count), which is evident from the great abundance of number systems in which the bases are the numbers 5 and 10. The first references to advanced and organised mathematics date back to the third millennium BC, in Babylon and Egypt. This mathematics was dominated by arithmetic, with some interest in measurement and geometric calculations and no measurement of mathematical concepts such as axioms and proofs. The earliest Egyptian books, written around 1800 BC, show a decimal numbering system with different symbols for successive powers of 10 (1,10,100,...) similar to the system used by the Romans. Numbers were represented by writing the symbol for 1 as many times as there were units in the given number, the symbol for 10 as many times as there were tens in the number and so on. To add numbers, the units, tens and hundreds...of each number were added separately. Multiplication was based on successive doubling and division was the reverse process (Encarta 2006).

The rapid development of microelectronics in recent years has made yesterday's results obsolete. Computers have become cheaper and their application possibilities have increased. Today, computers are not only used to solve large-scale numerical problems in technology and economics, but in a wide range of fields. The use of computers in all fields of human activity has made the development and application of science and technology in education, services and the economy and society in general even more rapid. The development achieved in contemporary science has promoted a series of transformations in all spheres of economic and social life. These transformations are also experienced in the field of education. The harmonious development of the personality of the new generations, the scientific conception of the world and the preparation of highly qualified professionals in accordance with the demands of the Scientific-Technical Revolution and the economic requirements of the country, demand the improvement of the quality of schools in general. The insertion of new information and communication technologies into the educational system from an early age is an essential part of the profound transformations that the Revolution is carrying out with great effort in this sphere with the aim of improving the quality of learning.

However, there are difficulties in addition with surplus in third grade because they do not respond with calculus skills, they lack skills in it. The insertion of educational software contributes to the achievement of these objectives, because through them the student interacts with information from different sources: texts, graphics, audio, videos, animations, photographs, tables, diagrams, maps and exercises. Today, a variety of educational software is available to the Cuban school which has all these resources, all of which combined make possible the development of general intellectual skills (observation, classification, comparison, evaluation) which are manifested in the increase of the processes of analysis, synthesis, abstraction and generalisation as a basis for thinking aimed at penetrating into the essence of the relationships between facts and phenomena. The demands made regarding the high degree of protagonism that the student must have within the teaching-learning process require a different conception of the role that the teacher must assume in its organisation and direction. These transformations must take place in terms of the conception, demands and organisation of the activity, as well as in the learning tasks conceived, thus achieving the participation of the learner and the search for and use of knowledge. The development of instructional activities in which exercises are used as a means of teaching implies having a broad knowledge of the contents that each one deals with and of all its possibilities. The entry of computers in education was not a casual occurrence, but a logical step in the development of the science that studies it and placed in the hands of pedagogues a powerful and versatile tool that can be exploited with great teaching success. For all these reasons, and taking into account the importance of education in the country as one of the fundamental achievements of the Cuban Revolution, all those responsible for carrying out the teaching-learning process at the different levels of the National Education System are making a greater effort every day to introduce changes aimed at achieving better results in the education of the new generations. For these reasons, the Minister of Education has outlined as one of his fundamental lines of work the demands made regarding the high degree of protagonism that the pupil should have in the teaching-learning process, which requires a different conception of the role that the teacher should assume in its organisation and management.

1.2 Theoretical foundation of the teaching-learning process of Mathematics in third grade.

In this chapter a historical analysis of the teaching-learning process of Mathematics is made, the mathematical calculation skills are characterised as well as the aspects that in some way influenced the development of the students' skills in the addition of mathematical calculation of addition with limit 10000 in Primary Education (EAC) and its influence on the teaching-learning process.

Mathematics is one of the oldest sciences whose development has been stimulated by the productive activity of men. As a particular science, with its own object of study, it has received the greatest influence from the natural sciences for the formation of new mathematical concepts and methods since its emergence. The teaching of mathematics in the education of third grade students consists in the fact that it not only prepares students for their incorporation into social and working life, but also contributes to the correction of their deficiencies. The teaching of mathematics develops the thinking of third grade students, achieving a progressive improvement in analysis, synthesis and generalisation. In the history of mathematics, there are many examples that show how the problems of the natural sciences were the genesis of important theories such as differential and integral calculus, which emerged as the most general method of solving mathematical problems, the theory of polynomials in relation to the investigation of the steam engine and many other cases can be cited, which show that mathematics is the result of the productive activity of men and that the new concepts and methods that make up their theories have had their roots in the fundamentals, in concrete problems of other sciences.

The peculiarity of the relationship of mathematics with other sciences, starting from the application of mathematical methods in the natural sciences, in the different periods of its development has been framed in two facets, as pointed out by K. Ribnikov (1987) in his book on the history of mathematics states:

- "The choice of the addition with mathematical surplus that corresponds approximately to the phenomenon or process, i.e. of the model, and the finding of the method of its solution".

- "The elaboration of new mathematical forms, since the approximation of the constructed mathematical model is inevitably imperfect. "1

This peculiarity in the application of mathematical methods up to the present day is evident in the development of cybernetics, computer technology, discrete mathematics, the growing role in economic, social and other sciences and the progress in it depends on the possibility of abstraction in the object of study and the choice of the logical scheme of abstract concepts representing the content of processes and phenomena.

For almost half of the last century, mathematics really had as its main research objective the metrical properties and relations in different types of magnitudes, it studied properties and relations of a mathematical nature, abstracting from their qualitative content, and was therefore qualified as a quantitative science.

Studies of the History of Mathematics, such as A. Aleskandrov (1980), in an effort to differentiate contemporary mathematics from the preceding one, emphasise its qualitative character, based on the broadening of its object and the deepening of the degree of knowledge of its objectives.

The passage of Modern Mathematics through the extensive use of the axiomatic method occurred after the discovery of non-Euclidean geometries and the appearance, at the end of the 19th century, of the theory of abstract sets with the axiomatic method led to the concept of abstract mathematical structure of sets created by G. Cantor.

The synthesis of theoretical ideas on set theory with the axiomatic method led to the concept of abstract mathematical structure which has been fundamental for all modern mathematics and which served as a premise for a group of French mathematicians (N. Bourbaki's group) to undertake the task of constructing existing mathematics on the basis of the concept of structure, considering this science, in its axiomatic form, as the accumulation of abstract forms which are applicable to a set of elements whose nature is not defined.

This transition to Modern Mathematics, characterised by a further growth in the levels of abstraction in mathematical levels and their relations, constitutes a qualitatively new step in the development of mathematical knowledge, which marks a qualitative and radical difference of current Mathematics with all that preceded it. The study of mathematical structures contributed greatly to the broadening of the field of application of modern mathematical methods, some of them such as the theory of groups and algebraic structures and functional analysis, which are expressions of the development and generalisation of concepts and ideas of classical mathematics, and others such as game theory and decision making, which respond to the needs of the social sciences. The mathematization of science is seen as a process of double growth of the concrete sciences and mathematics, which manifested itself in the emergence and successful development of sciences such as elementary particle physics, quantum chemistry, molecular biology and many others.

As a characteristic feature of the contemporary scientific-technical revolution, the increasing application of mathematical methods in the most diverse fields of science and technology necessitates a new understanding of the object and methods of contemporary mathematics. The content of the object of mathematics has been enriched in such a way that this has led to a restructuring and change in the totality of its important problems. We assume that the object of mathematics is enriched in inseparable relation with the requirements of technology and natural sciences, which is a necessary condition for understanding the place of this science in the productive and social activity of mankind, which does not reduce it only to an abstract science that studies quantitative relations and spatial forms far from reality. The understanding of the object of contemporary mathematics and its role in scientific and technical development leads to the analysis of what mathematics should be learnt, what a man needs in these times to face mathematical research, but, essentially, to face the wide diversity of other problems that require mathematical methods for their solution, from domestic problems to the most complex scientific problems. Mathematics comprises those factors that intervene and make it possible for mathematics to be taught and learnt, and in the last decade several authors have recognised the decisive influence of philosophical positions and epistemological theories on mathematical knowledge. In its broad sense, it is not restricted to the teacher-student interaction during the class, but goes beyond that to other factors that intervene in the teaching-learning process such as: "the design and development of study plans and programmes, textbooks, teaching methodologies, learning theories and the construction of theoretical frameworks for educational research, which are put into practice on the basis of the philosophical and epistemological conceptions that the teacher and students have about mathematics "2. The dominant philosophical conception of mathematics (formalist, realist, constructivist, etc.) has generated a type of mathematical activity at each stage of the development of this science, and a certain educational practice has been produced on the basis of this conception. In the formalist conception of mathematics, for example, which prevailed in the first half of this century, this discipline appears as a structured body of forms, alien to the meaning of objects. For its part, in the epistemological conception that understands the objects of Mathematics in a reality, which recognises its existence independent of the subject, based on the epistemological realism of Plato and Aristotle, it is proposed as a consequence that to know Mathematics is to recognise mathematical objects through processes of abstraction and generalisation in the corporeal objects of nature and under this conception, mathematical activity is close to the process of discovery of the mathematician.M. Santos (1990), points out, in this sense, that the learning of Mathematics is important the process and the sense that students show in the development or construction of mathematical ideas, he also points out that learning concepts about numbers, solving equations, graphing functions, etc., is not developing mathematics. "Doing or developing mathematics includes problem solving, abstracting, inventing, testing and making sense of mathematical ideas".3.

As indicated, knowledge, from the constructive perspective, is always contextual and is never separated from the subject, who is the one who assigns to the subject a series of meanings that conceptually determine the object.

Learning Mathematics has ceased to be understood as the simple accumulation of concepts, theorems or procedures of a certain order or relationship, which has led to this science being understood as something static, as a complex of terms and symbols that the student has to master.

In the analysis of new trends in the teaching of mathematics at the intermediate level, authors such as Panizza and Sadovki (1992) are of the opinion that "doing mathematics means elaborating definitions rather than repeating definitions given by others, it means looking for examples rather than asking for them, it means proposing counterexamples when one wants to demonstrate that a property is not valid, it means finding meaning in the hypotheses of a theorem, it means asking questions as well as answering them".

It is important in this position that they also recognise the need to take into account the deductive processes or activities that contribute to the modes of production and validation of mathematics as a formal science, which may be valid from the intermediate level of teaching, but it is questionable what place should be occupied by those processes of interaction that provide the necessary experiences to recognise not only the meaning of these concepts and examples, but also their applicability and existence in objective reality. Finally, in order to refer to contemporary trends in mathematics education, it is essential to quote Miguel de Guzmán (1992), who, based on the analysis of the main movements, transformations and results in the last decades, concludes that the current educational panorama of mathematics education in Spain is very different from that of the rest of the world, He concludes that the current educational panorama of Mathematics these general trends start from the question about the object of mathematical activity and from the clarification of what is the mathematical task and its influence on what should be the teaching of Mathematics, he assumes that mathematical activity is faced with a certain type of structures that lend themselves to some peculiar modes of treatment that include: adequate symbolisation, rigorous rational manipulation and effective mastery of the reality to which it is addressed.

1.3 Characterisation of the development of mathematical calculation skills in basic addition exercises with overshooting in third grade.

In the general mental development of the pupils, mathematical calculation exercises are important in terms of memorising basic addition exercises with overshooting.

An important step in the solution of this type of exercises is the memorisation of the basic ones up to limit 10 in the first grades, where the child begins to solve simple calculations that become more complex in the solution of exercises with overruns up to limit 20.

The work continues in third grade when the consolidation of calculation skills developed in addition and subtraction of natural numbers up to 1000 is worked on and the solution of exercises with natural numbers up to 1000 is initiated. It may be thought that here the pupil has all the skills to memorise and solve calculation exercises, but this is not the case, as new situations may arise that enrich the work with these exercises. An important contribution in new situations is made by the use of dynamic and practical exercises where the lesser student interacts by activating his or her psychic processes.

It is considered that the current state of third grade pupils' mastery in solving mathematical calculation exercises presents difficulties as they are not able to memorise and solve oral and written calculations on their own. Moreover, it is difficult for them to solve exercises.

Third grade pupils have difficulties in mathematical calculations, even when they have mastered the basic operations, so it is necessary to resort to special methods. It is important for pupils to memorise equalities, and for this purpose, the teacher can direct different activities that stimulate pupils to participate, one of which is the use of didactic games. It is known that in this grade of primary education, in relation to the age of the pupils, one of the fundamental activities is play. This is why it is necessary to link study with play or vice versa. These games, when well used and prepared with all their requirements, help pupils to assimilate, exercise and apply their knowledge, thus achieving the development of skills.

According to C. Rizo - (1985:42) in the related topic on the formation of skills and abilities in the teaching of Mathematics he states "that the development of skills in calculation are automated components of conscious activity. They arise through actions performed, first consciously, whose partial acts are grounded through frequent repetition and practice of the same activity until they become a unified act. The development of skills in calculation, abilities and knowledge are fundamentally integrated in the power of uniform performance".

We agree with this criterion in mathematics classes, specifically in exercise classes, where repeated exercises are included, as well as varied ones, including the solution of exercises with text, tables, equations and problems. In these classes, students must be kept motivated; the teacher must structure the class in a pleasant and interesting way, motivating the students' interest in learning. Among these activities, the most effective is the use of didactic games, which allow students to establish competitions, knowledge encounters, among others, thus enabling the development of mathematical skills in calculus. In the Methodology of Mathematics Teaching, according to S. Osvaldo (1991:107), skill is the ability of man to perform any action. The group of authors of the book Methodologies of Mathematics Teaching from first to fourth grade, Volume I (1976:184) defines ability as the actions that the subject must assimilate and therefore master from lower to higher degree and to this extent allows him/her to perform adequately in the relation of certain tasks.

An analysis of the above shows that there is no contradiction between them.

CHAPTER II: SYSTEM OF ACTIVITIES FOR THE DEVELOPMENT OF CALCULATION SKILLS IN BASIC ADDITION WITH OVERSHOOTING EXERCISES.

This chapter analyses the diagnosis of the current situation of the calculation of basic addition exercises with overrun in third grade students, and also shows the theoretical basis of the system of activities for the improvement of calculation skills in basic addition exercises with overrun in third grade students; finally, a validation of the experiment is carried out, showing the results achieved after applying the system of activities.

2.1 Diagnosis of the current situation of the development of calculation skills in basic addition exercises with overrun.

These instruments included the initial pedagogical test (appendix 1) applied to 17 pupils in the third year of the Reinaldo León Yera primary school with the aim of determining the current state of development of skills in the calculation of basic addition exercises with surplus through an oral calculation, where the results were not very flattering (appendix 2).

Of the 17 students, 11 passed (64.7%) and 6 failed (35.3%). When analysing the results achieved, only 11.8% (two students) achieved the category of excellent (E), they fully mastered all the basic addition exercises with overrun, they have developed all the necessary skills for calculation, they do not make mistakes; 17.6% (three students) obtained the grade of MB, they mastered the basic addition exercises with overrun, but they do not manage to do it quickly and accurately, they make mistakes in calculation.

23.5% (four pupils) reached the assessment of B, they have full mastery of only some of these basic exercises, the others are solved by means of auxiliary means such as fingers, they make mistakes in less than four calculations; 11.8% (two students) managed to calculate half of the exercises, recognise the operation correctly, use auxiliary means on several occasions to answer the calculation, are not quick or precise, and 35.3% (six students) do not manage to calculate less than half of the exercises, always use their fingers as a means to solve the exercise, sometimes use sets and confuse the operations.

An individual interview (annex 3) with the 17 third grade students with the aim of gathering information about how they develop the mathematical operations of addition with overrun, the route they use for the solution, as well as the level of development achieved in the skills of speed and precision, led to the following results.

88.8% (15 students) were able to recognise the addition sign with overshoot quickly and accurately, only 11.2% (two students) were not able to recognise the addition symbol accurately.

With regard to the route used by the child to calculate the given exercise, it was found that 47.1% (eight pupils) use their fingers as a way of solving the calculation, they have not achieved the skill of speed, 5.8%

(one pupil) use sets as a way of solving the calculation, and 47.1% (eight pupils) use the most suitable way for the calculation; memory, thus achieving the skills quickly and accurately.

In the calculation of basic addition exercises with overshooting, 11.8% (two pupils) respond quickly and accurately, so it can be deduced that most of the pupils do not memorise correctly all the basic addition exercises with overshooting, they do not have the necessary skills for full mastery of these exercises; 35.2% (six pupils) have to think a little and then respond to the calculation, so it is demonstrated that they have not developed the skill of speed, so precision is emphasised, 47.1% (eight pupils) need to use their fingers to carry out the calculation. This difficulty leads to the fact that the pupils have not developed the ability of speed and precision, they create habits and customs, the calculations are not conscious and only 5.8% (one pupil) is not able to master oral calculation, they need to make sets of balls or lines.

After having analysed the interview, it was found that more than 70% of the pupils do not manage to memorise the basic addition exercises with overshooting, i.e. they do not have the necessary skills for the satisfactory performance of this activity. All these difficulties are centred on the method of solution, as they do it with their fingers, by means of imagination, sets, etc. This has a negative influence on the development of skills with speed and precision, in these pupils it could be seen that they do not make an intellectual effort to analyse or reason about the solution of a given calculation exercise, they show that they feel more confident when they carry out the calculations with their fingers than when they do it consciously using their memory.

In terms of the time taken to solve an addition calculation exercise, 30.4% (5 students) calculated a basic addition exercise in less than three seconds and 70.6% (12 students) calculated a basic addition exercise in more than five seconds.

With regard to the effectiveness with which students carry out the calculations, it can be highlighted that only 11.8% (two students) do not manage to correctly solve half of the exercises proposed in class, 20.6% (three students) manage to carry out more than half of the exercises in class and 70.6% (12 students) are able to calculate consciously and accurately all the written calculations foreseen in the class.

Through the analysis of the observation of third grade students in mathematics classes, it was possible to determine that the greatest difficulties of the students in terms of mastering basic addition exercises with overshooting lie in the fact that the solution method used by most of the students is not the correct one, since they do not manage to develop the skills of speed and precision, they spend a lot of time calculating, they make this method a necessity, a habit, a habit.

2.2 Theoretical basis of the system of activities.

As a result of the difficulties that have been detected throughout the research, this section contains the theoretical foundations of the system of activities that will allow and raise the level of development of knowledge skills on the calculation of basic addition exercises with surplus in third grade pupils. It is proposed for a system of activities with the following characteristics.

- Integrative nature. It covers different components of mathematics (formal exercises, exercises with tables, with text and problems), and achieves an efficient development of calculation skills from the reproductive to the explanatory level.
- Flexible character. These activity systems can be used in teaching activities, specifically in exercise activities, they can be used as entertainment during break times, as well as in extracurricular and recreational activities.
- Guidance character. It serves as a methodological guide for the preparation of calculation activities on basic addition exercises with overlapping with third grade students.
- Competitive nature. To expand, assess the results achieved.

The system of activities consists of several regularities

- Be of adequate size to be observed by all pupils.
- Designing to attract the attention of learners
- Place in a visible location
- Link with educational aspects
- Participation of all pupils
- Be affordable according to grade and age

Some of the advantages the activity system offers you

- Most of the components of mathematics are worked on.
- Demands level of independence

- Helps to acquire knowledge and calculation skills of basic addition with overshoot exercises.
- They allow students to make assessments and self-assessments.

Content addressed in the system of activities.

- Addition of numbers
- Formation of basic exercise groups or pairs
- Recognition of terms
- Determining the parts and the whole
- The system of activities can be used in the exercise classes in the subject of Mathematics in third grade, fulfilling its purpose in unit #1 and theme 1.3 of the first period, these activities can be adapted to other contents of the different units, only that when elaborating the exercises it will be taken into account that these are adjusted to the objectives of the programme and to the unit that is being worked on.

The activity system consists of 20 activities, broken down as follows.

Activity 1 "The Dice of Calculation

Activity 2 "The calculus chain".

Activity 3 "Where I'm going".

Activity 4 "I am a sum".

Activity 5 "Shaping Equalities

Activity 6 "Calculando voy pintando" (Calculating as I paint)

Activity 7 "Who am I?

Activity 8 "How do I solve it?"

Activity 9 "Solve and Memorise".

Activity 10 "Calculate".

Activity 11 "Complete the values".

Activity 12 "I play and build

Activity 13 "I add and solve".

Activity 14 "Keep looking".

Activity 15 "My favourite calculation

Activity 16 "I learned to play".

Activity 17 "The Land of Numbers

Activity 18 "Happy Calculé" Activity 18 "Happy Calculé" Activity 18 "Happy Calculé" Activity 18 "Happy Calculé

Activity 19 "I found what was lost" Activity 19 "I found what was lost" Activity 19 "I found what was lost

Activity 20 "I solve problems".

Activities #1, 2, 3, 4, 5 and 6, 9, 10, 11, 12, 15, are based on the solution of formal exercises, very simple according to the content of each grade and require all the necessary data for their subsequent solution, they constitute the basis for the application exercises.

Activities #7, 16, 17, 18 and 19 make it possible to work with the exercises with text and tables where an unknown value or term is sought in the solution of an equality.

Activities # 8, 13, 14 and 20, allow the solution of problems where the parts and the whole are determined, solutions are given to different life situations, they offer data in a logical way that allows answering the question posed.

2.3 Methodological proposal of the system of activities

The methodological guidelines for implementing the set of exercises require the development of a system of activities. The following criteria of the authors consulted in the theoretical foundation were followed, which recognise the need to work on this content in close connection with the activities, taking into account the characteristics of the age of the pupils; the system of activities that can be used in Mathematics classes, in breaks, recreational and extracurricular activities is shown below.

Activity 1

Name: The dice of calculus.

Objective. To calculate orally and competitively in teams basic addition exercises with the development of dexterity skills, speed and precision when calculating, values of companionship, honesty and solidarity.

Contents: Formal exercises on the calculation of basic addition exercises with overshooting in third grade pupils in primary schools.

Activities to be carried out: The students must calculate orally, quickly and accurately the basic addition exercises with overshooting that would be formed when the dice are rolled, for the exercises the two dice are used, forming addition exercises such as.

9+4	**9+7**	**8+4**	**8+7**	**7+5**	**6+5**	**5+5**
9+5	**9+8**	**8+5**	**8+8**	**7+6**	**6+6**	
9+6	**9+9**	**8+6**	**7+4**	**7+7**	**6+7**	

In solving these exercises, they can be linked to numeracy content, taking into account the individual differences of the pupils, and other content can be integrated into the exercise.

Methodology: The classroom is divided into two groups, the dice are placed in a visible place for all the students in the classroom (these dice should have 8 to 10 cubic centimetres, have striking colours in the outline of the numbers), a student from the red team is sent to start the activity and the children from the other team continue doing it in turn. In order for the pupils to be able to solve the exercises, they are shown one.

Activity 2

Name: The calculus chain.

Objective: To calculate orally, quickly and accurately a chain of basic addition exercises using a system of activities that allows the development of mathematical skills.

Content: Calculation with basic addition exercises with overshooting in teaching and non-teaching activities of the subject Mathematics in the third grade of primary school.

Methodology: This activity can be developed by students, depending on the activity.

Activity 3

Name: Where I am going

Objective: To match the calculation of basic addition exercises with overshooting in a coherent and precise way through an activity that allows the development of calculation skills.

Content: Matching addition and surplus equalities with their respective results.

Activities to be developed: Motivation on the care and protection of flowers and butterflies. Linking the different calculation exercises with their corresponding results.

Methodology: Equalities are placed in each of the flowers with basic addition exercises with overlapping, and addition exercises that correspond to the equalities are placed in the butterflies.

Activity 4

Name: I am a sum

Aim: To recognise through different addition equals the corresponding sign.

Content: Form addition equals with the appropriate sign.

Activities to be undertaken: A brief discussion is held on the use of the signs representing the addition operation and their practical significance.

Methodology: A student will be given an activity that will match the first equality.

Activity 5

Name: Forming equalities

Aim: To form addition with overshoot equalities by calculating with the basic addition with overshoot exercises through an activity that enables the development of mathematical calculation skills.

Content: Formation of addition equalities from a given sum.

Activities to be carried out: Motivate with an activity such as the following:

Which basic exercises do I add 18?

Insist on when a number is placed in the geometric figure in the centre and it is greater than 10 means the formation of addition equalities with basic exercises.

Methodology: Show the students the activity, the teacher will place any number on the figure in the centre and the activity begins.

Activity 6

Name: Calculando voy pintando

Aim: To quickly and accurately calculate basic addition exercises with overshoot through an activity and in turn perform the calculation.

Content: Calculation of basic addition exercises with overshooting.

Activities to be carried out: Motivate students' knowledge of the different geometric figures (highlighting general characteristics) and the importance of using colours to bring things to life and make them colourful.

Methodology: Place in a visible place the silhouette of each of the drawings (house and boat) and place the geometric figures that make up these silhouettes on the table, where basic addition exercises with overlapping appear in each one. The activity begins when a student correctly answers the calculation that appears on the table, and then matches it with the silhouette in the drawing. The silhouettes must have the same number of elements to be painted.

Activity 7

Name: Who am I?

Aim: To find an unknown number or term for the behaviour of exercises with tables and with addition with overshooting texts through a system of activities that allows them to broaden their field of knowledge.

Content: Finding values and unknown terms in addition equals with overlapping through work with tables and exercises with texts.

Activities to be carried out: To motivate the students, they will be asked about the terms of the addition.

Methodology: The flannel board is placed in a visible place in the classroom and the elements required for the activity are inserted in it, whether they are exercises with tables (variables) or with texts (terms).

Activity 8

Name: How do I solve it?

Objective: To reason in the solution of a variety of simple problems with basic addition exercises with overshooting through a system of activities that enables the development of skills in calculus for practical application.

Activities to be developed: Motivate through the analysis of a simple problem all the necessary steps for the achievement of reasoning by the students. Explain the problem taking into account its parts and the whole.

Methodology: This system of activities can be applied in any mathematics class where there are exercises of this type. The exercise is presented to the students who are carrying out the activity, the students must represent the data provided by the problem with previously prepared figures, and then substitute them into the parts and the whole, asking themselves: What does it tell us? What are the parts? What is it that I want to find?

Activity 9

Name: Solve and memorise

Objective: To find an unknown number for the completion of basic addition exercises with overshooting using the activity system.

Content: Finding unknown values in addition equalities with overshoot using equality work.

Activities to be carried out: To motivate the students, they will be asked about the terms of the addition.

Methodology: A card is placed in a visible place in the classroom and the elements to be calculated are entered, whether they are exercises with two or three addends.

Activity 10

Name: Calcula

Objective: To quickly calculate the basic exercises of addition with surpassing through a system of activities where geometric elements are related to give colour to the drawing and at the same time to carry out the calculation.

Content: Calculation of basic addition exercises with overshooting related to known geometric figures.

Activities to be developed: Motivate students' knowledge of different geometric figures and their importance.

Methodology: Form rows of students and place in a visible place each of the drawings (flower-tree) and place on the table the geometric figures where the basic addition exercises with overlapping appear in each one.

Activity 11

Name: Fill in the vowels

Objective: To fill in the blanks with the vowels of the results obtained or terms for the completion of exercises with tables.

Content: Finding unknown terms in addition equals with surplus by working with tables.

Activities to be carried out: To motivate the students, the terms of addition will be presented to them in writing and then explained to them.

Methodology: This activity can be applied in any mathematics class where there are exercises of this type. The student is told that he/she is carrying out the activity and must complete it with the vowels he/she has made.

Activity 12

Name: Juego y Construyo

Objective: To calculate with agility and dexterity basic addition exercises with surpassing through a game where geometric elements that give colour to the drawing and at the same time carry out the calculation are related.

Content: Calculation of basic addition exercises with overshooting related to geometric figures.

Activities to be developed: Motivate students' knowledge of the different geometric figures (highlight their characteristics).

Methodology: Form teams and place in a visible place each of the drawings (flower) on the table geometric figures that make them up, where the basic exercises of addition with surpassing appear in each one. The game starts when the student of the first team takes one of the geometric elements; the calculation that appears in it is answered correctly and then it is made to correspond with the drawing, the drawing must have the same amount of elements when painted.

Activity 13

Name: I add and resolve

Aim: To solve basic addition exercises with overshooting skillfully.

Content: Solution of basic addition exercises with overshooting.

Activities to be carried out: Motivate students' knowledge of the different calculation operations in Mathematics.

Methodology: Form rows of students in the classroom and place on the floor geometric figures that make up the circle where basic addition exercises with overlapping appear in each of them. The game starts when a pupil in the first row picks up one of the geometric elements; the calculation that appears on it is answered correctly, then the pupil makes it correspond to the circle.

Activity 14

Name: Continue searching

Aim: To form addition equalities with overlapping by working with the known basic exercises through a system of activities that enables the development of mathematical calculation skills.

Content: Formation of addition equalities from a given sum.

Activities to be carried out: Motivate with an activity such as the following:

What basic exercises add 14?

Which basic exercises add 18?

Insist on when a geometric figure is placed in the centre of a number and this is greater than 10 means the formation of addition equalities with basic exercises.

Methodology: Divide the classroom into two teams, the teacher will place any number on the figure in the middle and the activity will begin.

Activity 15

Name: My favourite calculation

Aim: To recognise through different addition equalities the sign that corresponds to each one, taking into account the characteristics of the terms.

Content: Form addition equalities by matching the sign of each of the equalities formed.

Activities to be carried out: A discussion of the signs that represent each of the addition operations, their practical meaning, as well as the characteristics of each of the terms.

Methodology: One student will match the first equality with the corresponding sign, and so will the other students.

Activity 16

Name: I learned to play

Aim: To recognise through different addition equalities the sign that corresponds to each one, taking into account the terms.

Content: Calculate addition equalities by matching the sign of each of the equalities formed.

Activities to develop: Discuss the use of the signs that represent each of the addition operations and the relationship between the two, as well as their characteristics.

Methodology: The student must take equalities that represent the sum.

Activity 17

Name: The land of numbers

Objective: To accurately recognise basic addition exercises with overlapping through a map where elements that give colour are related and at the same time carry out the calculation.

Content: Recognition of the basic addition exercises with overlapping related to the map.

Activities to be developed: Motivate students' knowledge about maps (general characteristics), the importance and use of colouring.

Methodology: Form teams and place the map with its colouring in a visible place, place the figures that make up the map (jigsaw) on the table where basic addition exercises with overlapping appear in each of them, one student will start the activity and the others will continue.

Activity 18

Name: Feliz calculé

Aim: To form addition with overshoot equalities by working with the known basic addition with overshoot exercises through a system of activities that enables the development of mathematical calculation skills.

Content: Formation of addition equals from a given sum.

Activities to be carried out: Motivate with an activity such as the following:

What does this sign represent (+)?

Will it be added?

Insist on when we look at the addition sign that means the formation of addition equals with basic exercises.

Methodology: The teacher will place any number on the figure in the centre and the activity will begin.

Activity 19

Name: I found what was lost

Objective: To look for an unknown term or number in order to complete exercises with addition with overlapping texts through a system of activities that will allow them to extend their knowledge in memorisation.

Content: Finding unknown values or terms in addition equals with overshoot by working through exercises with texts.

Activities to be carried out: To motivate the students, they will be asked about the terms of the addition, which should then be explained.

Methodology: The classroom is divided into two groups and the flannel board is placed in a visible place in the classroom and the elements with which the activity is to be carried out are inserted, whether they are exercises with texts (terms).

Activity 20

Name: I solve problems

Objective: To reason in the solution of a variety of simple problems with basic addition with overshooting exercises through a system of activities that facilitates the development of calculation skills for their practical application.

Activities to develop: Motivate through an analysis of a simple problem all the necessary steps for the achievement of reasoning on the part of the students. Explain taking into account the parts and the whole.

Methodology: The system of activities can be applied in any mathematics class where there are exercises of this type, the exercise is given to the students who are executing the activity, and then they substitute in the parts of the exercise that they are executing.

2. 4 Evaluation of the effectiveness of the system of activities.

Of the 17 students, 16 passed for 94.12% and 1 failed for 5.88%.

When analysing the results achieved, only 24.9% (5 pupils) reached the category of excellent (E), they fully mastered all basic addition exercises with overshooting, they have created all the necessary skills for calculation, they do not make any mistakes; and 2 9.4% (5 pupils)

They mastered the basic addition with overshooting exercises, but they do not manage to do it quickly and accurately, they make mistakes in the calculation.

23.5% (4 students) reached the evaluation of good (B), they have full mastery of only some of these basic exercises, the others are solved by auxiliary means, they make mistakes in less than four calculations; 11.8% (2 students) managed to calculate half of the exercises; they recognise the operation correctly; 5.88% (1 pupil) do not manage to calculate less than half of the exercises, always use their fingers to solve the exercise, sometimes use sets and confuse the operations.

After analysing this pedagogical test, it was found that the greatest difficulties were in the way of solving the calculation and in the speed with which the exercises were solved, and it was also found that the basic addition exercises were difficult to solve with overruns.

Conclusions of the Chapter

This chapter provides detailed information on the diagnosis of the current situation of the problem through the application of various measurement instruments when evaluated qualitatively and quantitatively, as well as the proposal of a system of activities, its theoretical basis and its effectiveness when applied to third grade students.

CONCLUSIONS

The characterisation of the basic addition exercises with overshooting made it possible to point out that the ability to memorise is of great importance in the teaching-learning process of mathematics, as it is the basis for the written procedures in later grades.

The teaching-learning process in mathematics can be improved through a system of activities aimed at developing calculation skills in basic addition exercises with overrun.

BIBLIOGRAPHY

ALBARRÁN PEDROSO, JUANA. Didactico de la Matemática en la Escuela Primaria. Ciudad de La Habana. Ed. Pueblo y Educación, 2005.248p.

ÁLVAREZ PÉREZ, MARTHA. Tratamiento de la aritmética. - Pedagogía 2001. -Havana. -- 2001 (booklet).

ARISTOS. Diccionario ilustrado de la Lengua Española. Havana: Ed Pueblo y Educación, 1974. -- 692p.

BALLESTER PEDROSO, SERGIO. Metodología de la Enseñanza de la Matemática. Tomo I. Ciudad de La Habana. Ed. Pueblo y Educación, 1992. 459p.

BALLESTER PEDROSO, SERGIO. Metodología de la Enseñanza de la Matemática. Tomo II. Ciudad de La Habana. Ed. Pueblo y Educación, 2002. 336p.

BERNABEU PLOUS, MATILDE. Alternative to develop and train the ability of calculus in the first place. -- Master's thesis -- ISP Enrique José Varona: Ciudad de la Habana, 1998. --187p.

BERNABEU PLOUS, MATILDE. Cuaderno Complementario Matemática tercer grado. Havana City. Ed. Pueblo y Educación, 2005. 37p.

BRITO, HÉCTOR. Psicología General para Institutos Superiores pedagógicos: Ed Pueblo y Educación: Ciudad de La Habana, 1987. - 213p.

CARRERO GONZÁLEZ, NITZA. Juegos Didácticos y Capacitación Profesional. Havana: Ed Pueblo y Educación, 1996. - 85p

CASANOBA, FRANSISCO. A structuring of the teaching and learning of numeracy and calculation in the first grades of primary school. Master's thesis. - ISP Raúl Gómez García: Guantánamo, 2001. -121p.

CASTILLO, MAYELÍN. The treatment of calculus operations in first grade based on the part-whole relationship. - Diploma thesis. - Manuel Ascunce Domenech School: Ciego de Avila, 2001. 129p.

CASTILLÓN LOXALA, JOSÉ ANTONIO. Playing to learn: Ciego de Ávila -- Marcelo Salodo Primary School, 1999. --23p.

FONSECA VELIZ, MARÍA ELENA. Calculation skills in first and second year primary school pupils. Master's thesis. --Ciego de Avila, 2004. -- 132p.

FUNLABRADA, IRMA. Play and learn Mathematics. Activities for fun and work in the classroom: Mexico, 1991. -- 96p.

GARCÍA BATISTA, GILBERTO. Fundamentos del Investigación Educativa. Primera parte. Havana City. Ed. Pueblo y Educación, 2001.15.

GEISLER CASTRO, VICENTE. Metodología de la Enseñanza de la Matemática de primer a cuarto grado. Havana: Pueblo y Educación, 1998. -- 179p.

GÓMEZ OLIVERA, EVANGELI0. Una concepción metodológica para la enseñanza de la Matemática en segundo grado. Evento Internacional Pedagogía 97: Ciudad de la Habana, 1997 (Paper).

GUTÍRRES MORENO, RODOLFO. Los componentes de proceso pedagógico y su dinámica. --ISP Félix Varela. --Villa Clara 1997. -(magnetic support).

H, WUSSING. Lecture on History of Mathematics: Ed Pueblo y Educación. Ciudad de la Habana. --282p. Indicaciones a los maestros de primaria para lograr habilidades de cálculo. -School year, 1986/1987. --20p. Hacia un Didáctica desarrollada: Ciudad de la Habana. -- Ed. Pueblo y Educación, 2001. --115p-

INCHÁUSTEGUI VILLALÓN, MIRIAN. Mathematics third grade. Ciudad de la Habana. Ed. Pueblo y Educación, 1990. --173p.

JUNGK, W. Conferencia sobre Metodología de la Enseñanza de la Matemática 1. Havana, 1978. --180p.

LABARRERE, GUILLERMINA, Pedagogía: Ed. Pueblo y Educación, 1998. --354p.

LÓPEZ HURTADO, JOSEFINA. El carácter científico de la pedagogía en Cuba: Ed. Pueblo y Educación. La Habana, 1996. --226p.

MARTÍ PÉREZ, JOSÉ. Educación Científica en los escritos sobre Educación. Havana: Ed. Pueblo y Educación, 2001. --170p.

MATIENZO PALMA, ADELAIDA. Orientaciones Metodológicas tercer grado. Volume I. Havana City. Ed. Pueblo y Educación, 1990. --208p. Metodología de la Enseñanza de la Matemática de primero a cuarto grado. Collective of authors of RDA. Havana City: Ed. Pueblo y Educación, 1978. --255p.

MIRANDA TORRES, SABINA. Orientaciones Metodológicas de segundo grado. Havana: Ed. Pueblo y Educación, 2001. --170p.

NOCEDO de LEÓN, IRMA. Metodología de la Investigación educacional. Primera Parte. Ciudad de la Habana. Ed. Pueblo y Educación, 2002. --192p.

PÉREZ MARTÍN, DALIA. Research work on the teaching of mathematics. --José Martí Primary School. Ciego de Avila, 1986. --30p. (Booklet).

PÉREZ SAMPER, EMILIA. Second grade programme. La Habana: : Ed. Pueblo y Educación, 2001. --91p.

PÉREZ SOMOZA, ELPIDIO. Metodología de la Aritmética Elemental. Havana: Pueblo y Educación, 1930. --305p.

PÉREZ RODRÍGUEZ, GASTÓN. Metodología de la Investigación Educacional. Havana: Ed. Pueblo y Educación, 1986. --139p.

PÉREZ RODRÍGUEZ, GASTÓN. Metodología de la Investigación Educacional. Primera parte. Havana City: Ed. Pueblo y Educación, 2001. --139p.

PENA GALVES, ROSA LINDA. Orientaciones Metodológicas de segundo grado: Ciudad de la Habana: Ed. Pueblo y Educación, 1989. -- tomo2.

PITA, BALBINA. El tratamiento del cálculo en el primer ciclo. Havana: Ed. Pueblo y Educación, 1985. --81p.

RICO MONTERO, PILAR. Hacia el perfeccionamiento de la escuela primaria. Havana City: Ed. Pueblo y Educación, 2000. --137p.

RIZO CABRERA, CELIA. National Seminar, 2001. --tomoII.

RODRÍGUEZ IZQUIERDO, JESÚS. Third Grade Programme. Volume I. Havana City: Ed. Pueblo y Educación, 1990. --94p.

RODRÍGUEZ RODRÍGUEZ, GERMÁN. Alternative for advantageous calculus in the first and second grades of primary school --Master's Thesis. Havana City, 1998, --115p.

SIMÓN, OSVLADO. Metodología de la Enseñanza de la Matemática en la escuela primaria. Havana: Ed. Pueblo y Educación, 1991. --209p.

SMAKARENKO, ANTÓN. Lecture on children's education. Havana: Ed. Pueblo y Educación, 1974. --121p.

VILLALÓN, INCHÁVSTEGUI. La elaboración de ejercicios básicos en la enseñanza de la matemática. Havana: Revista Educación. --July-September, 1978. --127p.

ZHUKÓUSKALA R, I. El juego y su importancia pedagógica: La Habana: Ed. Pueblo y Educación, 1982. --140p.

Printed by Books on Demand GmbH, Norderstedt / Germany